INTRODUCTORY CHEMISTRY MATH REVIEW TOOLKIT

GARY L. LONG
SHARON D. LONG
Virginia Tech

Upper Saddle River, NJ 07458

Assistant Editors: Jessica Neumann and Carol DuPont
Editor-in-Chief, Science: Nicole Folchetti
Senior Editor: Kent Porter Hamann
Assistant Managing Editor, Science: Gina M. Cheselka
Project Manager, Science: Ed Thomas
Supplement Cover Manager: Paul Gourhan
Supplement Cover Designer: Victoria Colotta
Operations Specialist: Amanda A. Smith

Pearson Prentice Hall
Pearson Education, Inc.
Upper Saddle River, NJ 07458

Printed in the United States of America

11 12 13 14 15 16 V036 15 14

ISBN-13: 978-0-13-601858-2

ISBN-10: 0-13-601858-0

Pearson Education Ltd., *London*
Pearson Education Australia Pty. Ltd., *Sydney*
Pearson Education Singapore, Pte. Ltd.
Pearson Education North Asia Ltd., *Hong Kong*
Pearson Education Canada, Inc., *Toronto*
Pearson Educación de Mexico, S.A. de C.V.
Pearson Education—Japan, *Tokyo*
Pearson Education Malaysia, Pte. Ltd.

Contents

Acknowledgements

The authors (GLL and SDL) would like to acknowledge the contributions of Timothy Smith and Diane Vukovich on Section 1 of this book.

Preface — To The Student

A successful study of general chemistry requires you to use your skills of memorization, logic, and mathematics. From my twenty-five years in the college classroom, I have found that students who are unsure about their math skills generally do poorly on their exams. To put this statement into context, imagine that you are holding a chemistry exam containing 20 questions, with 15 of them involving the use of mathematical equations. If you have trouble with the first few math questions and have only an hour to complete the exam, any weakness in math skills will limit your performance on the exam, regardless of how many facts and figures you have memorized.

The math skills that a student needs to successfully complete general chemistry are basically what you learned in high school: algebra and trigonometry. Calculus is not required. It is the purpose of this booklet to guide you in the mathematics that are used in your first year of General Chemistry.

The case of a life sciences student planted the idea for this booklet. This young man was going around for the seventh time in general chemistry. He could never manage to pass the first test so he would drop out and wait until the next semester to try again. Fortunately, he came to me at the beginning of the course. He was desperate; he could not graduate until he passed chemistry. After speaking with him, I found his limitation was not inadequate math training, but a basic fear of math problems. He would stumble over the math problems on the exam. Over the course of the semester I met with the student and explained how to perform these calculations using the methods that are described in this booklet. The student successfully completed the course with a B grade.

Although we cannot absolutely guarantee your success in chemistry with the use of this booklet, mastery of the material presented here will enhance your ability to work problems in general chemistry. Take a few moments and complete the "Self Test for Math Skills" at the back of this booklet. It will help you assess your skills. Use this guide to help you reinforce your math skills.

Chemistry is an exciting science that touches every aspect of our lives. It is our hope that this booklet will demystify the mathematics used in general chemistry and help you discover this excitement.

Gary L. Long
long@vt.edu

I will never forget my own struggle with chemistry at Wake Forest University. As a freshman, I had not yet developed productive study habits. I faced every new chemistry chapter with a certain dread. There is a tendency to procrastinate when we fear or find something distasteful, or so it was in my case with chemistry. However, if you cope with chemistry in this manner, as I did, the stress of catching up and preparing for a test will cost you your well being until the test is over. The two grades of C that I made in freshman chemistry taught me this painful lesson. The point is that even if you dislike chemistry and are taking it only to satisfy the requirements for your major, make your effort consistent.

Many college students have been forced to change their career dreams because they could not make it through chemistry. This unpleasantness does not give this science a good name nor does it give chemistry professors an easy grace at social functions when asked their occupation. Some of you may need to change your career option to a different discipline for which you are gifted. A very kind English professor pointed this out to me at a time when organic chemistry was causing me great distress. Time is too valuable to spend it in the wrong place. However, for those of you heading in the right direction, we do not want chemistry to be a stumbling block for your dreams.

We hope that your experience with chemistry will be a victorious one and that in some small way this math book will make a difference. Write us and tell us your stories as well as your triumphs.

Sharon D. Long
sdlong@vt.edu

Harnessing the Power of Your Calculator

The best friend you have in a chemistry class is your calculator. If used correctly, it will save you precious seconds during a test, make doing homework problems outside class much more efficient, and allow you to spend more time on the concepts rather than laboring over the calculations.

In your general chemistry course, you will encounter problems using arithmetic operations (addition, subtraction, multiplication, and division), trigonometric operations (cos, sin, $\cos^{-1}$, and $\sin^{-1}$), and logarithmic operations (log, ln, 10^x, e^x, powers, and roots). You also will deal with scientific notation in many of these problems. While your textbook has an excellent appendix on mathematics for general chemistry, in this toolkit, we present methods for using your calculator to solve these calculations.

Take the time to acquaint yourself with the capabilities of your calculator. Please try the practice problems in Section 1.12 of this toolkit. You will find your calculator to be a powerful tool that you can harness for your study of general chemistry.

1.1 TEN-DOLLAR CALCULATORS THAT WORK

Many students have difficulty in the first few weeks of the semester because they don't have the type of calculator they need in order to quickly and efficiently perform the tasks the instructor and textbook will require of them. You may sit down in your first class with a four-function calculator that just does the basic operations of adding, subtracting, multiplying, and dividing (the type the bank gives you when you open a checking account). This type of calculator is not powerful enough to do many of your chemistry problems. On the other hand, you may not want to purchase an engineering calculator that has so many functions it is intimidating to look at and has a manual as thick as *Harry Potter and the Order of the Phoenix*.

You need a scientific calculator that will help you quickly and easily dispatch the problems set before you in your class, especially on tests. The calculator should have the following capabilities:

- the four basic arithmetic functions;
- a button to enter scientific notation (EE or EXP on most calculators);
- buttons to easily enter logarithms;
- buttons to convert between scientific notation and decimal form;
- buttons to do statistical calculations (you will not need them in this class, but many of you will probably take statistics later in your program);
- at least 10 places on the display so you can display fairly large numbers in decimal form (called 10 + 2 on some models).

Below is a list of calculators that are inexpensive (around $10) and do the job you need a calculator to do

- Casio FX-260
- Texas Instruments TI-30Xa or TI-30X IIS
- Sharp EL-501

1.2 The First Five Buttons —The Basic Operations

Nothing is easier than multiplying or dividing on a calculator. If you want to multiply 2.71 × 1.91, you simply punch

2.71 [×] 1.91 [=] The display reads 5.1761

When you divide two numbers, the process is the same. Let's divide 3.75 by 1.2. This problem can be presented to you as 3.75 [÷]

1.2 or 3.75/1.2 or $\frac{3.75}{1.2}$. No matter how the division is written, it is entered into the calculator as

3.75 [⍰] 1.2 [=]   The display reads 3.125

Most of the time in chemistry you will do several operations in one problem. An example of this type of problem is

$$2.875 \times \frac{1}{3.125} =$$

Many students enter the operations like this.

2.875 [×] 1 [⍰] 3.125 [=]   The display reads 0.92

It is much simpler and less prone to error if the multiplication step is left out when all you are doing is multiplying by one. You simply enter

2.875 [⍰] 3.125 [=]   The display reads 0.92

Later, when you do conversions, you will do many multiplications and divisions in one problem. Let's try one.

$$\frac{2.375 \times 4.15}{0.03125 \times 1.577} =$$

Many students do the problem as follows.First, they multiply the numbers in the numerator:

2.375 [×] 4.15 [=]   The display reads 9.85625

Then they write this answer down, push [C] to clear the display, and then multiply the numbers in the denominator:

0.03125 [×] 1.577 [=]

The display reads 0.04928125

Then they write this number down, clear the display, reenter 9.85625, and finally divide by 0.04928125 to get the answer 200.

In other words, they multiply the numbers in the numerator first, erase the answer, then multiply the numbers in the denominator, erase the answer, reenter the first answer, and then do the division.

There is an easier way to solve this equation. Do the multiplications and divisions in order from left to right. Press [×] before numbers that are in a numerator (after the first number) and press [÷] before any number in a denominator.

2.375 [×] 4.15 [÷] 0.03125 [÷] 1.577 [=]

The display reads 200

You can change the order, as long as you press [×] before a number in the numerator and press [÷] before a number in the denominator.

2.375 [÷] 0.03125 [÷] 1.577 [×] 4.15 [=]

The display reads 200

Let's try another.

$$1.44 \times \frac{1}{16} \times \frac{13.3}{0.20} =$$

We enter:

1.44 [÷] 1.6 [×] 13.3 [÷] 0.20 [=]

The display reads 5.985

Doing problems this way can save you much time on tests.

1.3 Rounding Off and Significant Figures

If you divide 2 by 7 on your calculator, the display reads 0.285714286. Ten decimal places can be quite cumbersome. Fortunately, you do not have to write all those numbers when you answer a problem; you can round off your answer. Answers beyond a certain number of places are not only cumbersome, but they are misleading. So we round our answers off to a reasonable number of decimal places.

In order to round off, we have to understand significant figures (sometimes referred to as sig figs). In chemistry (and in all science), we do not work with pure numbers as in algebra class. We work with measurements. If you take your body temperature with a fever thermometer, it may look like the figure below. The mercury stops between 98 and 99 degrees. It looks like it is between 98.2 and 98.4 degrees (each small line on the thermometer is 0.2 of 1 degree), so we might guess 98.3. In all measurements, the last digit is a good guess and your guess might be a little different from mine. This temperature (98.3 degrees) has three significant figures. If we round it off to two places, it will have two significant figures (98 degrees).

A thermometer showing a reading of 98.3 degrees.

Some laboratory balances read to the hundredths place. The digital readout of an object's mass might look like this: 14.49. This reading has four significant figures. Other (more expensive) balances read to the ten-thousandths place. The digital display on this balance might read 14.4927 for this object. This reading has six significant figures.

1.4 What Are the Rules for Significant Figures?

How can you tell if digits are significant? Most of the confusion involves the number zero. Your textbook gives you rules similar to these.

All non-zero digits are significant.

This means that any number that is not zero is significant. 7.121486 has seven significant figures; 4215 has four significant figures.

All zeros between non-zero digits are significant.

7002 has four significant figures; 7.1002 has five significant figures; 702 has three significant figures; 80.003 has five significant figures.

All zeros to the left of the first non-zero digit are not significant.

Here is where your normal, everyday definition of "significant" may clash with the chemistry definition. The number 0.0025 has only two significant figures, since the zeros to the left of the two are not significant. That does not mean they do not have to be there. If you took them away, you would get .25, a very different number! For instance, a common dosage of a blood pressure-lowering drug is 0.0025 grams. If you ignored the zeros, the dosage would become .25 grams, 100 times too high and probably fatal to the patient. Those zeros have the very important job of being placeholders; however, they are not significant in the sense of expressing the accuracy of the number.

If we convert 0.0025 kilograms to grams, it becomes 2.5 grams. Because we changed the units, we no longer needed the left-hand zeros as placeholders.

When a decimal point is present, zeros to the right of the last non-zero digit are significant.

Thus, 0.2300 has four significant figures; 2.4900 has five significant figures; 2.0 has two significant figures.

If there is no decimal point written, you cannot tell if the zeros to the right are significant. This rule can confuse us if we are not careful.

For instance, we cannot say how many significant figures whole numbers such as 8000 and 400 have. They have one significant figure; beyond that, we cannot say.

Now let's look at some examples that involve all the rules:

- 7.0010 has five significant figures. (All the zeros are significant because two are between non-zero digits and the last is in a number that has a decimal point.)
- 0.0005 has one significant figure. (The zeros to the left are not significant.)
- 0.000500 has three significant figures. (The zeros to the left of the 5 are placeholders, but are not significant. The zeros to the right are significant because a decimal point is shown.)
- 1.000500 has seven significant figures. (All the zeros are significant because the three zeros are between non-zero digits and the last two are to the right in a number that has a decimal point.)
- 7000 has one significant figure. (No decimal point is shown, so you cannot tell whether the zeros are significant.)
- 7000. has four significant figures. (This time the decimal point is shown.)

1.5 Scientific Notation to the Rescue

Let's look at the number 800. Suppose it represents 800 grams. If 800 grams is just a good guess at the weight of an object, then the number has only one significant figure. However, if you weighed the object and it weighed exactly 800 grams, the number would have three significant figures. If you want to show that the number has three significant figures, you would have to write it with a decimal point included; that is, 800. grams. But it is easy for a decimal point at the end of a whole number to "get lost." However, we can easily show the number of significant figures in numbers such as this by converting them to scientific notation (also called exponential notation). If you convert 800 to scientific notation, it becomes 8×10^2.

If you want to show that it has three significant figures, you can write it as 8.00×10^2.

When you write a number in scientific notation, the number of significant digits is expressed by the coefficient.

- 7000 grams becomes 7×10^3 with one significant figure
- 7000 grams becomes 7.0×10^3 with two significant figures
- 7000 grams becomes 7.00×10^3 with three significant figures
- 7000 grams becomes 7.000×10^3 with four significant figures

Remember that the exponent plays no part in determining the number of significant figures. It merely tells us where to put the decimal point.

1.6 Rules for Rounding Off

You may remember these rules from a math class in high school. To round off a number to a certain place, you look at the number in the next place to the right of the number. If that number to the right is less than 5 (that is, if it is 0, 1, 2, 3, or 4) you leave the number alone. This is called rounding down. If the number to its right is 5 or greater (that is, if it is 5, 6, 7, 8, or 9) you increase the value of the number by one. This is called rounding up.

Let's round some numbers.

- 7.134 rounded to the hundredths place becomes 7.13 since the number in the thousandths place is 4, a number less than 5.
- 0.21029 rounded to the ten-thousandths place becomes 0.2103 since the number to the right of that place is 9, a number that is 5 or greater.

1.7 Significant Figures in Multiplication and Division

When you multiply or divide measurements, the number with the least number of significant figures limits your answer. Suppose three students measure a box. Jane measures the length to be 3.12 cm, Carl measures the width to be 3.1 cm, and Donna measures the height to be 7.62 cm. To get the volume, we multiply the three sides together.

3.12 cm [×] 3.1 cm [×] 7.62 cm [=]
The display reads 73.70064, with the answer in cm^3

However, we must round the calculated answer to 74 cm^3 because Carl made his measurement to only two significant figures.

When we calculate density, we divide mass by volume. That is,

$$\text{Density} = \frac{\text{Mass}}{\text{Volume}}$$

Let's say we measure a sample's mass to be 7.4219 grams and its volume to be 1.23 mL. We divide 7.4219 by 1. 23.

$$\text{Density} = \frac{7.4219\text{ g}}{1.23\text{ mL}}$$

$$\text{Density} = 6.034065041\ \text{g}/\text{mL}$$

Because 1.23 mL has only three significant figures, we must round the answer to 6.03 g/mL.

Significant figures are important, but do not get too bogged down with them. They will come naturally to you as you do more calculations.

1.8 [EXP]: Scientific Notation on the Calculator

Calculators make scientific notation easy. You just need to have the right calculator and know which buttons to push.

Note: Scientific calculators have a button that is marked EE (on most Texas Instruments calculators) or EXP (on most other brands). This is the button that lets you enter a number in scientific notation.

Let's take the number 7.45×10^8. To enter this number, we enter the coefficient 7.45, then press [EXP]. Two zeros appear at the right (smaller in size). These zeros are holding the places for the two digits of the exponent. (You will never need more than two.) We then enter the exponent 8. The display should look like this: ⍰7⍰5 08

When you see this display, you read it as 7.45×10^8. Press [C] to clear the display, and let's enter it again.

7.45 [EXP] 8 The display reads 7.45 08

Note: A common mistake students make is to hit [X] before [EXP]. Do not do this! The "times 10 to the" is all included in [EXP].

So you *do not* enter

7.45 [×] [EXP] 8 or 7.4501 [×] 10 [EXP] 8

Simply press 7.45 [EXP] 8

Another example is to enter 9.12×10^{12}

Press 9.12 [EXP] 12 The display reads 9.12 12

For an entry with negative exponent, try 7.216×10^{-9}

Enter 7.216 [EXP] 9 [±] or 7.216 [EXP] [±] 9
The display reads 7.216 -09

The [±] key toggles a number between negative and positive. If you press [±] after [EE] or [EXP], it toggles the exponent between negative and positive. Press [±] a couple of times and see how the sign of the exponent changes back and forth.

Note: The [±] key is not the same as [–]. The [±] toggles between negative and positive. The [–] key is used only for subtraction.

Let's enter a few more numbers in scientific notation. Enter the mass of an electron, which is 1.6606×10^{-24} kilograms.

Press 1.6606 [EXP] 24 [±]

The display reads 1.6606 $^{-24}$

The second example is to enter 10^5. On your calculator:

Press 1 [EXP] 5 The display reads 1.0 ^{05}

Note: When entering an exponential number where no coefficient is shown, you must supply the missing coefficient of 1. We can write 10^5 as 1×10^5.

Finally, let's enter 10^{-7}.

Press 1 [EXP] 7 [±] The display reads 1.0 $^{-07}$

1.9 MULTIPLICATION AND DIVISION WITH SCIENTIFIC NOTATION

If you multiply two numbers in scientific notation, just press [×] between the numbers. Let's multiply $(2.12 \times 10^4)\ (7.89 \times 10^{-9}) =$

Enter 2.12 [EXP] 4 [×] 7.89 [EXP] 9 [±] [=]

The display reads 1.67268 $^{-04}$

Since each number has three significant figures, we round the answer to 1.67×10^{-4}. Note that the only time you enter [×] is between the two numbers.

Let's divide two numbers: $\dfrac{8.81 \times 10^4}{7.472 \times 10^8}$

Enter 8.81 [EXP] 4 [⍰] 7.472 [EXP] 8 [=]

The display reads 1.1790685 $^{-04}$

Since the least number of significant figures is three, we round our answer to 1.18×10^{-4}.

Let's try another: $\dfrac{10^{-14}}{10^{-9}}$ (Remember to supply the missing coefficients of 1.)

Press 1 [EXP] 14 [±] [□] 1 [EXP] 9 [±] [=]
The display reads 1.0000000 $^{-05}$, written as 1×10^{-5} or 10^{-5}.

Now let's look at one last example that uses Avogadro's number (6.02×10^{23}), a number that you will see later in the class.

$$2.50 \times 10^{15}\left(\frac{12.0}{6.02 \times 10^{23}}\right)$$

Press 2.5 [EXP] 15 [x] 12 [□] 6.02 [EXP] 23 [=]
The display reads 4.□833887 $^{-08}$

Since each of the three numbers has three significant figures, we round the answer to 4.98×10^{-8}.

1.10 Switching Between Scientific and Decimal Notation

Sometimes you may want to convert an answer back and forth between scientific notation and decimal form. This will be most important to you on exams, especially multiple-choice exams where the answer choices could be written in either form.

The answers to problems in the back of your textbook may also be given in either scientific notation or decimal form. Often students are confused when they look up a hard-fought answer in the answer key and discover that their answer looks wrong. Their answer may not be wrong. The textbook may have given the answer in scientific notation, while the student calculated it in decimal form.

Texas Instrument Calculators

On Texas Instrument calculators, we use [FLO] and [SCI] along with [2nd] to convert between scientific notation and decimal form. On the TI-30X calculator, [FLO] and [SCI] are second functions written in blue above [4] and [5]. We use [2nd] to access these operations.

Pushing [2nd] [FLO] puts the calculator into normal decimal mode. (FLO comes from "floating point," computer programmers' jargon for a decimal number.) In this mode, any number we enter will change to decimal form as soon as we hit another key (as long as the number will fit the display).

Pushing [2nd] [SCI] puts the calculator into scientific notation mode. Any number we enter will change to scientific notation as soon as we hit another key.

First let's change some numbers in scientific notation to decimal form. Suppose you got 7.7×10^{-4} as an answer and want to convert it to decimal form.

Enter 7.7 [EE] 4 [±] The display reads 7.7^{-04}

Press [2nd] [FLO] The display reads 0.00077

Now let's convert 8.449×10^{7} to decimal form.

Enter 8.449 [EE] 7 The display reads 8.449^{07}

Press [2nd] [FLO] The display reads 84490000

Now let's try to convert 6.419×10^{-17} to decimal form.

Enter 6.419 [EE] 17 [±] The display reads 6.419^{-17}

Press [2nd] [FLO] The display still reads 6.419^{-17}

Nothing happens because a number in scientific notation with an exponent of –17 has too many decimal places to fit into the 10 spaces on the calculator. A number in scientific notation must have an exponent of 9 or less, either plus or minus, to fit into the 10 decimal places on a calculator. That is, a number must be between nine billion and nine-billionth.

Now let's take some decimal numbers and convert them to scientific notation. Suppose you got 0.000788 for an answer and want to change it to scientific notation.

Enter 0.000788 — The display reads ⍰.000788

Press [2nd] [SCI] — The display reads 7.88 $^{-04}$

Convert 6,992,000 to scientific notation.

Enter 6,992,000 — The display reads 6992000

Press [2nd] [SCI] — The display reads 6.992 ^{06}

You can press [2nd] [SCI] to convert to normal decimal form.

Note: If you press [2nd] [SCI] to convert an answer to scientific notation, the calculator remains in that mode until you press [2nd] [FLO] to put the calculator into normal decimal mode. If you enter a decimal number, it will change to scientific notation as soon as you press another function key such as [×], [÷], or [=] .

Likewise, if you are in decimal mode when you enter a number in scientific notation, it will change to decimal form as soon as you hit another key (as long as the decimal form will fit in the display). In either of these cases, the number is not affected, but only the form of the number.

Let's look at a few examples. Press [⍰⍰⍰] [⍰CI] to put the calculator in scientific notation mode.

Enter 0.0059 The display reads ⍰⍰⍰59

Press [×] or [⍰]

The display changes to 5.9000000^{-03}

Press [C] to clear the display. Now press [2nd] [FLO] to change your calculator to normal decimal mode.

Now enter 0.0059 The display reads 0.0059

Now press [×] or [⍰] The display still reads 0.0059

Casio, Sharp, and Most of the Rest

Most brands of calculators have a [MODE] button, usually in the upper left hand corner next to [INV]. They also have a table printed under the readout display that looks something like this:

MODE	•	0	4	5	6	7	8	9
	SD	COMP	DEG	RAD	GRA	FIX	SCI	NORM

or this:

MODE	0:COMP	4:DEG	5:RAD	6:GRA
	• SD	7:FIX	8:SCI	9:NORM

The two numbers you need in this table to convert between scientific notation and decimals are 8:SCI (scientific notation) and 9:NORM (for normal decimal form).

Pressing [MODE] [9] puts the calculator into normal decimal mode. In this mode, a number you enter will change to decimal form as soon as you hit another key (as long as the number will fit in the display).

Pressing [MODE] [8] puts the calculator into scientific notation mode. Any number you enter will change to scientific notation as soon as you hit another key. The second number, the one you enter after pressing [MODE] [8], tells the calculator how many places of the coefficient the calculator will display. For instance, if you press [MODE] [8] [4], the answer will be rounded to four digits. It's better to let the calculator display eight digits; then you can choose the proper number of significant figures for an answer.

Let's change some numbers from scientific notation to normal decimal form. Suppose you got 7.7×10^{-4} as an answer and want to change it to decimal form.

Enter 7.7 [EXP] 4 [±] The display reads 7.7^{-04}

Press [MODE] 9 The display reads 0.00077

Convert 2.568×10^{8} to decimal form.

Enter 2.568 [EXP] 8 The display reads 2.568^{08}

Press [MODE] 9 The display reads 256800000

Now let's try to convert 7.489×10^{-13} to decimal form.

Enter 7.489 [EXP] 13[±]

The display reads 7.489^{-13}

Press [MODE] 9 The display still reads 7.489^{-13}

Nothing happens because a scientific notation number with a –13 exponent has too many decimal places to fit into the 10 spaces on the calculator. A number in scientific notation must have an exponent of 9 or less, either plus or minus, to fit into the 10 decimal places on a calculator. That is, a number must be between nine billion and nine-billionth.

Now let's convert decimal numbers to scientific notation. Suppose you got 0.00489 as an answer and want to change it to scientific notation.

Enter 0.00489 The display reads 0.00489

Press [MODE] 8 8

The display reads 4.8900000^{-03}

Convert 2,090,000 to scientific notation.

Enter 2,090,000 The display reads 2090000

Press [MODE] 8 8 The display reads 2.090000^{06}

Press [MODE] 8 8 to put the calculator in scientific notation mode.

Note: If you press [MODE] 8 8 to convert an answer to scientific notation, the calculator remains in that mode until you press [MODE] 9 to change to normal decimal mode. If you enter a decimal number, it will change to scientific notation as soon as you press another function key such as [x] or [÷].

Likewise, if you are in decimal mode when you enter a number in scientific notation, it will change to decimal form as soon as you press another key (as long as the decimal form will fit in the display). In either of these cases, the number is not affected, but only the form of the number.

Let's look at some examples. Press [MODE] 8 8; this puts the calculator in scientific notation mode.

Enter 0.0059 The display reads 0.0059

Press [×] or [?]

The display changes to 5.9000000^{-03}

Press [C] to clear the display. Now press [MODE] 9 to return your calculator to normal decimal mode.

Now enter 0.0059	The display reads 0.0059
Now press [×] or [?]	The display still reads 0.0059

1.11 Does the Answer Make Sense?

A calculator is a very handy instrument. It will make computations quick and easy, but it is only as good as the numbers you enter. Computer programmers have an acronym for this: GIGO. Garbage in, garbage out. If you are in a hurry on a test, it is easy to push the wrong button and not notice it. Let's look at some calculations and see if we can tell if the answer makes sense.

$$\frac{9.2\times10^{8}}{5.5\times10^{5}}=1.7\times10^{3}$$

We can check the answer to see if it's in the right ballpark by looking at the exponents: $10^8 - 10^5 = 10^3$. Also notice that 9.2 is a little less than twice 5.5, so the ratio of 1.7 is reasonable. Notice that the numerator is larger than the denominator, so the answer has a positive exponent.

Let's look at another example:

$$\frac{7.9\times10^{-4}}{6.9\times10^{-9}}=1.1\times10^{5}$$

We look at the exponents: $10^{-4} \div 10^{-9} = 10^{-4-(-9)} = 10^{-4+9} = 10^5$. It checks out. In this case, we divided a larger number by a smaller number and we get a number greater than one. Also notice that 7.9 is just a little larger than 6.9, so a ratio of 1.1 makes sense.

If we had hit the [×] button instead of the [÷] button by mistake, we would have gotten the wrong answer 5.5×10^{-12}. By quickly checking the exponents, we can spot this kind of error. The few seconds you spend checking your answers is time well spent.

Remember that studies show that most students can raise their grades one or even two letters by doing nothing more than being neat and checking their work.

Look at the following problem and the incorrect answer given.

$$\frac{7.50 \times 10^9}{\left(6.5 \times 10^7\right) \times \left(3.5 \times 10^3\right)} = 4.0 \times 10^7 \quad ???$$

Do a quick check of the exponents. Did your check tell you that the exponent in the answer should be around 10 and not 10^7 (that is, $10^9 \div 10^5 \div 10^3 = 10^{(9-5-3)} = 10^1$)?

Can you figure out what the student did wrong to arrive at the incorrect answer? The student carried out the calculation *incorrectly* as:

7.5 [EXP] 9 ÷ 6.5 [EXP] 5 × 3.5 [EXP] 3 [=]

instead of the correct key sequence:

7.5 [EXP] 9 ÷ 6.5 [EXP] 5 ÷ 3.5 [EXP] 3 [=]

Remember that on previous pages we noted that all numbers in the denominator must be preceded by ÷ even if a × sign appears in front of the number. This student got the wrong answer because she forgot this rule.

On multiple-choice tests, the wrong answer you are likely to get is often one of the choices. Let's look at one last example.

$$12 \times \frac{1}{1{,}000} = 12{,}000 \quad ???$$

Here the student probably pressed [×] instead of [÷]. Since 1,000 is in the denominator, it will make 12 one thousand times smaller instead of one thousand times larger.

1.12 PRACTICE PROBLEMS

Use your calculator. Round all answers to the appropriate number of significant figures. Answers are on the next page.

1. $\dfrac{4.807 \times 1.85}{0.467 \times 5.00} =$

2. $\dfrac{96.8}{15.2 \times 0.0120} =$

3. $10.7 \times \dfrac{1.2}{3.8} \times \dfrac{14.2}{5.01} =$

4. $\left(4.16 \times 10^{5}\right) \times \left(5.08 \times 10^{-8}\right) =$

5. $\dfrac{7.00 \times 10^{-6}}{5.871 \times 10^{4}} =$

6. $\dfrac{5.68 \times 10^{3}}{5.02 \times 10^{2}} =$

7. $\left(10^{5}\right)\left(10^{-8}\right)\left(10^{3}\right) =$

8. $\dfrac{10^{-14}}{10^{-9}} =$

9. $\left(3.15 \times 10^{18}\right) \times \left(\dfrac{342}{6.02 \times 10^{23}}\right) =$

10. $\left(1.00 \times 10^{15}\right) \times \left(\dfrac{1.66 \times 10^{-24}}{12.0}\right) =$

1.13 Answers to Practice Problems

1. 4.807 [×] 1.85 [÷] 0.467 [÷] 5 [?] 3.808543897

Rounded to 3 sig figs, the answer is 3.81

2. 96.8 [÷] 15.2 [÷] 0.012 [=] 530.7017544

Rounded to 3 sig figs, the answer is 531

3. 10.7 [×] 1.2 [÷] 3.8 [×] 14.2 [÷] 5.01 [=] 9.577056413

Rounded to 2 sig figs, the answer is 9.6

4. 4.16 [EXP] 5 [×] 5.08 [EXP] 8 [±] [=] 0.0211328

Rounded to 3 sig figs, the answer is 0.0211

In scientific notation, the answer is 2.11×10^{-2}

5. 7 [EXP] 6 [±] [÷] 5.871 [EXP] 4 [=] 1.1923011^{-10}

Rounded to 3 sig figs, the answer is 1.19×10^{-10}

6. 5.68 [EXP] 3 [÷] 5.02 [EXP] 2 [=] 11.31474104

Rounded to 3 sig figs, the answer is 11.3

In scientific notation, the answer is 1.13×10^{1}

7. 1 [EXP] 5 [×] 1 [EXP] 8 [±] [×] 1 [EXP] 3 [=] 1

The answer is simply 1

8. 1 [EXP] 14 [±] [÷] 1 [EXP] 9 [±] [=] 0.00001

The answer in scientific notation is 1×10^{-5}

9. 3.15 [EXP] 18 [×] 342 [÷] 6.02 [EXP] 23 [=] 1.7895349^{-03}

Rounded to 3 sig figs, the answer is 1.79×10^{-3}

The decimal form is 0.00179

10. 1 [EXP] 15 [×] 1.66 [EXP] 24 [±] [÷] 12 [=] 1.38333^{-10}

Rounded to 3 sig figs, the answer is 1.38×10^{-10}

Scientific Measurements

2.1 Significant Digits

Our calculators normally display six to nine digits when performing arithmetic functions. While it may seem advantageous to have such large numbers when performing chemical computations, not all of this information may be significant because of error associated with the data.

To put this into context, consider the data that you record in the laboratory. Every measurement made is subject to error. The level of this error (or certainty) depends on the instruments used in the measurement and the skill of the person performing the operations. Even if we can eliminate the systematic errors (*i.e.*, miscalibration of the pan balance) we will still encounter random errors in the lab. These random errors will determine the accuracy of our measurement.

Rather than stating the error with each number a chemist may use in a calculation, the practice of using "significant digits" in calculations is employed. The significant digits of a number can be thought of as those digits in the number that do not change when the uncertainty is factored in. (The rules used for determining the number of significant digits in an expressed value are found in your textbook.) As an example of the effect of uncertainty on the number of significant digits, consider a food sample, which was found to have 1.33 mg of Ca per serving with an error of 0.1 mg for the determination. Based on the uncertainty, we should only say that the sample contains 1.3 mg of Ca. The number has only two significant digits. It is of no importance to attempt to convey that the calculation indicated a second 3 in the decimal. The error made this digit "insignificant." For practice in significant digits and rounding off nonsignificant digits, try Self-Test Problems 2 and 3.

You will be asked in this course to use significant digits in the estimation of answers to your calculations. As an example, consider the problem of determining the circumference of a cylinder. With a simple ruler you could determine the diameter to be 2.5 inches. From the relationship of $C = \pi \square d$, you could use your calculator to determine the circumference, C. The display of the calculator would read (7.853981…). What value do you report? What is significant?

The basic rule is that you report your answer using the least number of significant digits. In the multiplicative operation above, the diameter was only reported to two significant digits. The answer should be reported as 7.9 in. See Self-Test Problem 5.

Rules for addition and subtraction are slightly different. Consider adding a 0.001 g weight to an object resting on a pan balance and indicating a value of 23.2 g in mass. The addition of 0.001 to 23.2 would yield a theoretical answer of 23.201 g. However, if the pan balance used had an error of 0.1 g, the addition of 0.001 g to this weight would be insignificant. The uncertainty will limit our ability to express accurately the sum of the combined weights. See Self-Test Problem 4 for work with this concept.

2.2 Scientific Notation

Scientific Notation (also called Exponential Notation) is a method used to express very large and very small numbers on your calculator. These numbers are expressed by multiplying the significant portion of the number by a multiplier based on 10^x.

To use the scientific notation feature on your calculator, you will use a [SCI], [EXP] or [EE] function. Consider the process of entering $1.6 \square 10^{-17}$. First enter the significant portion of 1.6 into the display. Next press the [SCI] or equivalent key and enter the number 17. You will need to change the sign of the power of 17 (which should appear in the right hand portion of the display), so press the [+/-] key to cause the power to read −17. Depending on the type of calculator you have, you may need to press the [=] key to end the sequence.

You will also be able to convert from regular notation to scientific notation on your calculator. If the number 1,000,000 was in the display of your calculator, you could depress the [SCI] then [=] keys to cause the number to appear in scientific notation (1.000 □ 10^6) on the display.

An example of scientific notation covered in the chapter on atoms in your text may involve the determination of the frequency of a photon from its wavelength. For instance, if a HeNe laser has a wavelength of 632.8 nm, what is the frequency of the photon? The answer is calculated using $c = \nu\lambda$, where c is the speed of light (a constant), ν is the frequency (in Hertz) and λ is the wavelength in meters. We can quickly estimate the frequency by expressing all numbers in scientific notation and setting up the problem. Using this, we can write:

c = 2.998 □ 10^8 m/s
λ = 632.8 nm or 632.8 □ 10^{-9}m

or 6.328 □ 10^{-7} m

The problem can be set up to solve for ν as:

$$□ = \frac{c}{\lambda}$$

$$\nu = \frac{2.998 □ 10^8 m/s}{6.328 □ 10^{-7} m}$$

$$\nu = 4.738 □ 10^{14} Hz$$

Before using the calculator, we can get a rough idea of the magnitude of the number. The exponential portion of the calculation shows a +8 in the numerator and a –7 in the denominator. Using our algebra skills, we can determine the magnitude of the exponent is +15. The non-exponential term of the equation can be reduced to 3/6, which is 1/2 (or 0.5). Hence, our rough guess of the answer is 0.5 □ 10^{15}Hz, or 5 □ 10^{14}Hz.

The calculator would allow us to find the answer of 4.738 □ 10^{14}Hz, based on 4 significant digits. This answer is close to our estimate. More examples of scientific notation can be found in Self-Test Problems 6 and 7.

2.3 MEAN AND STANDARD DEVIATION

As your text discusses, every measurement in the laboroatory has some level of uncertainty. This uncertainty is often not evident from just looking at the data. Instead, the uncertainty is often calculated based upon all the data you have available. Often, these calculations involved the determination of the mean and standard deviation of the data. These two values are helpful in determining the accuracy and precision of your measurements.

If your calculator has a statistical analysis package, you may be able to calculate the mean ($\bar{x}$) and standard deviation (s) of a data set by simply entering the data into the calculator and pressing the appropriate function key. As an example, let us say you titrated five equal volumes of an unknown acid with a standard base. You have dispensed the following volume of titrant in the 5 runs: 43.80 mL, 43.30 mL, 44.10 mL, 43.90 mL and 43.70 mL. We would first enter the five values into the memory of the calculator. After the final entry, we would depress the button to give us the mean, $\bar{x}$. The value of 43.76 should appear. Next, we would calculate the standard deviation, s, by pressing the appropriate buttons (s, s_x, or σ). A value of 0.30 should appear.

Based on your titrating skills, you could say you dispensed 43.76 +/- 0.30 mL per titration in this data set. Considering the error in the data (0.30 mL), expressing a value beyond the tenths of mL (0.1 mL) would not be significant. The correct answer should be 43.8 ± 0.3 mL, where only three significant digits are used.

The Metric System

Goals: To perform calculations with specific heat.
Skills: Multiplication and division.

The concept of specific heat is outlined in your text in the chapter on energy. This term allows you to calculate the heat required to raise the temperature of one gram of a specific element, compound, or material. Conversely, this term will allow you to determine the heat that is released or absorbed by a chemical reaction if the temperature change of the reaction can be measured.

As an example of this concept, consider the heat that is generated when solutions of acids and bases are mixed together. The heat evolved in a reaction can be expressed using the following equation:

$$\frac{\text{heat (cal)}}{\text{mass (g)} \times \text{temperature change } (\Delta t)} = \text{specific heat}\left(\frac{\text{cal}}{\text{g} \times {}^{\circ}C}\right)$$

The mass for the mixture is determined by weighing or measuring the volume of liquid in the calorimeter and multiplying this volume by the density of the solution. The specific heat is normally given in a table in your text. The Δt is calculated from the temperature readings. The Δt term is defined as the final temperature minus the initial temperature. A reaction where heat evolved (exothermic) will have a positive Δt while a reaction where heat is absorbed from the environment (endothermic) will have a negative Δt.

Let us say that when the acid and base were mixed in the calorimeter, the temperature of the solution changed from 23.5°C to 25.9°C. The final volume of the mixture was 50.0 mL. Since the density of the solution (being mostly water) is 1.00 g/mL, the aqueous solution weighs 50.0 g. The specific heat of water is 1.00 cal/(g × °C).

The Δt is calculated as 25.9 - 23.5°C = 2.4°C. Placing these values into the above equation yields:

$$\frac{\text{heat (cal)}}{50.0 \text{ g} \times 2.4\ ^{\circ}\text{C}} = 1.00 \left(\frac{\text{cal}}{\text{g} \times {}^{\circ}\text{C}} \right)$$

$$\text{heat (cal)} = 120 \text{ cal}$$

See Self-Test Problem 10 for work with this type of expression.

A point to remember is that any expression that involves reactions will deal with the difference in the final state from the initial state. For any calculation using Δ (*i.e.*, Δt) make a habit of thinking that Δ = (final state – initial state). Using this approach you won't get the sign of the Δ wrong.

Models of the Atom

Goals: To calculate the wavelength, frequency and energy of photons.
Skills: Multiplication, division and scientific notation.

In your study of radiant energy, you will investigate the relationship of the wavelength and frequency of photons. The basic equation is:

$$c = \nu \times \lambda$$

where ν is the frequency of the photon (expressed in Hertz) and λ is the wavelength of the photon (expressed in m). The speed of light, c, is a constant and is approximately 3.00×10^8 m/s. Care must be taken in this calculation to keep all values in the correct units. For instance, λ of a Na vapor lamp is 589 nm. To determine ν, we can use the above equation. However, all values involving a measure of length in this calculation need to be expressed in the same units. It is most convenient to express them in meters for this problem. The wavelength is in nm, so the value would be expressed as 589×10^{-9} m, or better yet, 5.89×10^{-7} m. By rearranging the above equation to solve for ν, we write:

$$\nu = c / \lambda$$
$$\nu = (3.00 \times 10^8 \text{ m/sec}) / 5.89 \times 10^{-7}\text{m}$$
$$\nu = 5.09 \times 10^{14}\text{/sec or Hz}$$

This conversion of wavelength to frequency is very useful in dealing with problems in the quantum theory section where the energy of a photon is related to the frequency of the electromagnetic radiation using $\Delta E = h\nu$. You can use this relationship also for working the energy of a photon based on its wavelength.

Stoichiometry

Goals: To calculate yields.
Skills: Multiplication and division.

Part of your studies dealing with chemicals and their reactivity will be to determine the yield of a reaction. From your readings in text concerning stoichiometry, if we write the equation:

$$A + B \rightarrow C$$

we understand if 1 mol of A and 1 mol of B are allowed to react under favorable conditions, 1 mol of C should be produced. The *theoretical yield* of the reaction is 1 mol of C. This value could be expressed in grams instead of moles by multiplying the theoretical yield of 1 mol by the molar mass of the compound C.

Note that yield depends on the stoichiometry of the reaction. If the reaction were instead:

$$2\,A \rightarrow B$$

and 1 mol of A was placed in the reaction vessel, the theoretical yield would be 0.5 mol of B. Using the unit analysis problem-solving method (outlined in your text) the problem could be solved.

$$x \text{ mol B} = 1 \text{ mol A} \times \frac{1 \text{ mol B}}{2 \text{ mol A}}$$

$$x \text{ mol B} = 0.5 \text{ mol B}$$

In practice, the mol quantity of product that is obtained from the reaction is less than the theoretical value. The amount of product that is obtained is termed the *actual yield.*

A method in which the efficiency of a reaction is gauged is the calculation of the *percent yield.* It is defined as:

$$\text{percent yield} = \frac{\text{actual yield}}{\text{theoretical yield}} \times 100\%$$

For the first reaction, if we had obtained 0.8 mol of C, the % yield would be:

$$\text{percent yield} = \frac{0.8 \text{ mol}}{1 \text{ mol}} \times 100\%$$

$$\text{percent yield} = 80\%$$

The calculation of the percent yield can also be performed using the grams of actual product obtained as compared to the theoretical gram yield. See Self-Test Problem 9 for more work with percent yield.

In many cases, the scientist may wish to predict the actual yield for a well-characterized reaction. For instance, if the percent yield of a synthetic reaction is 90%, the actual yield may be found by multiplying the percent yield by the theoretical yield. (Remember to divide the percent yield by 100 in order to remove the % unit from the expression.)

The Gaseous State

Goals: To perform calculations with gas laws and the ideal gas law.
Skills: Cross multiplication and division.

6.1 Gas Laws

In this section of your studies, you will be exploring relationships between pressure, *P*, volume, *V*, temperature, *T*, and mol quantities of gases, *n*. Your text will introduce you to Boyle's law, Charles' law, Gay-Lussac's law, and the combined gas law.

You must exercise care in two areas to avoid mistakes when using these laws. The first area is temperature: all temperatures must be expressed in Kelvin (K) even if the data is given to you in °C. Remember, the relationship is K = °C + 273.15. The second area is to rearrange the equation so that the left side only contains the term that you are seeking. All the given data would then appear on the right side of the equation.

Consider the use of Charles' law in finding the volume that a gas in a balloon would occupy if it was heated from 25°C to 75°C. The initial volume of the gas (V_1) at 25°C is 1.00 L. First, let's set up the equation:

$$\frac{V_1}{T_1} = \frac{V_2}{T_2}$$

From the prior information we can write:

$V_1 = 1.00$ L
$T_1 = 25°C$, which is 298 K
$V_2 = ?$
$T_2 = 75°C$, which is 348 K

To solve for V_2, we write:

$$\frac{V_1 T_2}{T_1} = V_2$$ (cross multiplying T_2)

$$V_2 = \frac{V_1 T_2}{T_1}$$ (switching sides)

By plugging in the data, we obtain:

$$V_2 = \frac{1.00 \text{ L} \times 348\text{K}}{298 \text{ K}}$$

$V_2 = 1.17$ L

When dealing with Boyle's, Charles', Gay-Lussac's law, or the combined gas law, you will often find a great deal of information in the problem. Follow these steps to solve such problems:

1) Write the gas law equation to be used.
2) Sort the given data (as set 1 or set 2).
3) Rearrange the equation to solve for the unknown term.
4) Enter the data and solve.

See Self-Test Problems 4-7 for more work with the mathematics of these gas laws.

6.2 PARTIAL PRESSURES

The pressure of a gaseous mixture is dependent on the partial pressure exerted by each gas. This relationship is known as Dalton's Law of Partial Pressures and is expressed as $P_1 = X_1 P_t$, where P_t is the pressure of gas_1, X_1 is the mole fraction of gas_1, and P_t is the total pressure of the gas. To illustrate this law, consider a container where 1.0 mol of gas_1 is mixed with 3.0 mols of gas_2. The mole fraction of gas_1 is calculated as follows:

$$X_1 = \frac{1.0 \text{ mol of gas}_1}{4.0 \text{ mols of total gas}}$$

$$X_1 = 0.25$$

similarly,

$$X_2 = \frac{3.0 \text{ mols of gas}_2}{4.0 \text{ mols of total gas}}$$

$$X_2 = 0.75$$

IMPORTANT: The sum of all the mole fractions should always add up to 1. The mole fraction is a simple ratio; it has no units.

Now, to work the problem, let's assume the pressure in the container was measured to be 2.00 atm. What is the partial pressure of gas_1? The equation is:

$$P_1 = X_1 P_t$$

The terms to use in the equation are:

$X_1 = 0.25$ (calculated above)

$P_t = 2.00$ atm (given in the problem)

Now by substituting into the equation we obtain:

$$P_1 = 0.25 \times 2.00 \text{ atm}$$

$$P_1 = 0.50 \text{ atm}$$

By similar reasoning we could determine that $P_2 = 1.50$ atm. Also, since the gas mixture consists of only two gases, if the pressure of gas_1 is 0.50 atm and the total pressure of *both* gases is 2.00 atm, by difference we can find the pressure of gas_2 to be 1.50 atm.

The use of Dalton's law, to correct for the true pressure of a gas collected over water, is the subject of the math in Problem 9 of the Self-Test section.

6.3 Ideal Gas Law

Another law that you will use is the ideal gas law, $PV = nRT$. Here, n is the number of moles of the gas and R is the gas constant [(0.0821 L·atm)/(K·mol)]. This gas law will allow you to work any of the previous problems that relate the P, V, and T of a gas to the mole quantity of the gas. For instance, what volume would 1.00 g of CO_2 occupy at 1.00 atm of pressure and a temperature of 23°C? Begin this problem by writing the equation.

$$PV = nRT$$

We were not given n in the original problem, but we can calculate it by using the equation $n = g/molar\ mass$. Here,

$$n = \frac{1.00 \text{ g}}{44.0 \text{ g / mol } CO_2}$$

and is equal to 0.0227 mol. With this information we can write:

$P = 1.00$ atm
$V = ?$
$n = 0.0227$ mol
$R = 0.0821$ (L·atm)/(K·mol)
$T = 23°C$, which must be expressed as 296 K

Rearranging the equation to solve for V, we write:

$$V = \frac{nRT}{P}$$

Plugging in the data:

$$V = \frac{0.0227 \text{ mol} \times 0.0821 \text{ (L}\cdot\text{atm)/(K}\cdot\text{mol)} \times 296 \text{ K}}{1.00 \text{ atm}}$$

$$V = 0.552 \text{ L}$$

Another important use of the ideal gas law involves the determination of the molar mass of a gas as well as the mass of the gas in grams. The molar mass (MM) of a gas is the weight of one mole of gas. It can be determined by replacing the n term in the ideal gas law with g/MM, where g is the grams of gas.

Solutions

Goals: To calculate concentration and dilutions.
Skills: Multiplication and division.

7.1 CONCENTRATION UNITS

A very important factor of solutions involves concentration units. In the chapter on solutions in your text, you are introduced to two terms: mass percent and molarity. Both of these terms involve the measurements of the mass of the solute and the amount of solution.

The mass percent is a simple ratio of the mass of solute to the mass of the solution. The term "percent" involves the multiplication of the simple ratio by 100%. Both masses are obtained by weighing the solute and the solvent. For instance, if we place 10 g of NaCl in a beaker with 100 g of water, we can say the mass of the solute is 10 g and the mass of the solvent is 100 g.

The calculation of the mass percent involves placing the mass of the solute in the numerator of the equation and the mass of the solution in the denominator. Since our solution contains 100 g of water and 10 g of NaCl, the mass of the entire solution is 110 g. This is shown below.

$$\text{mass/percent} = \left(\frac{10\text{ g of NaCl}}{100\text{ g of water} + 10\text{ g of NaCl}}\right) \times 100\%$$

$$\text{mass/percent} = \left(\frac{10\text{ g of solute}}{110\text{ g of solution}}\right) \times 100\%$$

$$\text{mass/percent} = 9.1\%$$

The mathematics of these calculations involve addition and division. See Problems 8 and 9 of the Self-Test for work with this concept.

The molarity, M, is the term that expresses the mol quantity of a dissolved substance (solute) per liter of solution. A solution that is 1.00 M NaCl contains 1.00 mol of NaCl per 1.00 L of solution. This same solution contains 58.4 g of NaCl per 1.00 L of solution.

Calculations that involve molarity will usually begin with the statement of the gram quantities of solute dissolved in a quantity (mL or L) of solution. By definition,

$$M = \frac{\text{mol of solute}}{\text{L of solution}} = \frac{\left(\dfrac{\text{g of solute}}{\text{MW of solute}}\right)}{\text{L of solution}}$$

For instance, to calculate the molarity of a solution that is 10.0 g of NaCl in 400 mL of water, we would write:

$$M = \frac{\left(\dfrac{10.0\text{ g}}{58.4\text{ g/mol}}\right)}{0.400\text{ L}}$$

We will first solve the numerator of the problem for the mol amount of NaCl present in the solution. If we divide 10.0 g by 58.4 g/mol, we should obtain a value of 0.171 mol NaCl. Substituting into the equation

$$M = \frac{0.171\text{ mol}}{0.400\text{ L}}$$

$$M = 0.428\text{ mol NaCl / L of solution}$$

Note: It is important to only express the volume of solution in L.

You may find it necessary in your work to determine what amount of solute, expressed in grams, would result in a certain molarity of solution. Consider what gram amount of NaCl is necessary to create a 0.60 M solution that is 750 mL in volume. First, let's fill the given information into the molarity expression.

$$0.60\ \text{M} = \frac{\left(\dfrac{x\ \text{g}}{58.4\ \text{g/mol}}\right)}{0.750\ \text{L}}$$

To solve for x, we must cross multiply several factors in order to arrange the equation for x.

$$(0.60\ \text{M})\ (0.750\ \text{L}) = \boxed{\frac{x\ \text{g}}{58.4\ \text{g/mol}}}$$

$$(0.60\ \text{mol/L})\ (0.750\ \text{L})\ (58.4\ \text{g/mol}) = x\ \text{g}$$

$$x\ \text{g} = 26.3\ \text{g of NaCl}$$

Note that all units on the left side of the equation reduced to g (remember that M is in mol/L). By taking time with your cross multiplication step, the problem can be solved.

7.2. Dilution

Dilution involves reducing the concentration of a solution by adding more solvent. Typically, a volume of solution is taken and added to a volume of solvent. Consider the dilution of a 3.00 M NaCl solution by pipetting 10.0 mL into a 1.000 L flask and filling the flask to the calibration mark with pure water. Here we are adding roughly 990 mL of water to the solution. The concentration of the diluted solution is calculated as:

$$M_f = M_i \times \frac{V_i}{V_f}$$

where M_f is the final concentration of the diluted solution, M_i is the initial concentration of the concentrated solution, V_f is the final concentration of the diluted solution, and V_i is the initial concentration of the solution. For our problem, M_i = 3.00 M NaCl, V_i = 10.0 mL, and V_f = 1.000 L (or 1,000 mL). Substituting these numbers into the equation, we write:

$$M_f = 3.00 \text{ M} \times \frac{10.0 \text{ mL}}{1{,}000 \text{ mL}}$$

$$M_f = 0.0300\ M$$

Note that the units of volume in a dilution calculation must be the same no matter how they are stated in the problem. Use the unit analysis problem-solving method to check your units before reporting your answers.

The dilution equation can also be used to solve for V_i, if the other variables of the equation are known. For instance, you may be asked to find what volume (V_i) of a 16.0 M solution of HNO_3 (M_i) must be added to the beaker and diluted to 2.25 L (V_f) in order to produce a diluted solution of 0.100 M HNO_3 (M_f). Rearrangement of the dilution equation by cross multiplication and division yields:

$$V_i = V_f \times \frac{M_f}{M_i}$$

from which V_i can be found as 0.0141 L (or 14.1 mL). Compare this example to Problem 10 of the Self-Test.

Acids and Bases

Goals: Calculations of acid-base titrations and pH.
Skills: Multiplication, division, logarithms and scientific notation.

8.1 TITRATION

Another calculation for solutions that involves cross multiplication and division is titration. As discussed in your textbook, this process involves reacting a standard solution with an unknown solution for the purpose of determining the concentration of the unknown solution. In the lab, commonly used titrations involve acid-base reactions, precipitation reactions, redox reactions, and complexation reactions.

Before any titration can be carried out, the solution stoichiometry must be known. Specifically, we must know how many moles of the titrant (standard solution) will react with the unknown. Using a simple acid-base reaction that is listed below, the mathematics for a titration can be set up.

$$HCl(aq) + NaOH(aq) \rightarrow H_2O(l) + NaCl(aq)$$

This reaction has a stoichiometry of 1 mol of acid to 1 mol of base. We can then write:

$$mol_{acid} = mol_{base}$$

Substituting the expression of mol = M × V, we obtain

$$M_{acid} \times V_{acid} = M_{base} \times V_{base}$$

Consider a titration of an unknown NaOH solution with a standardized HCl solution. To 50.0 mL of the unknown contained in a flask, approximately 33.8 mL of a 0.175 M standardized HCl solution was added to the flask to reach the equivalence point. From this information, we can determine the molarity of the unknown solution.

If we rearrange the above equation by cross multiplication and division, we can solve for the molarity of the unknown base, M_{base}. Doing so yields:

$$M_{base} = \frac{M_{acid} \times V_{acid}}{V_{base}}$$

Inserting the data (with volumes in L), we find :

$$M_{base} = \frac{0.175 \text{ mol/L} \times 0.0338 \text{ L}}{0.0500 \text{ L}}$$

$$M_{base} = 0.118 \text{ M}$$

Please note that the reaction in this example possesses a solution stoichiometry of 1:1.

For an example where the solution stoichiometry is different, see your text. If an acid-base reaction were to use H_2SO_4 and NaOH, The diprotic acid would react with 2 moles of the NaOH base. The solution to this type of titration problem requires starting out with the equation:

$$1 \text{ mol}_{acid} = 2 \text{ mol}_{base}$$

From this stoichiometric relationship, the equation to find the M_{base} using H_2SO_4 as the standardized acid in this titration can be developed just as we have seen above. More problems using ratios are found in Problems and 6 of the Self-Test. By using the methods described above, these problems on acid-base standardization can be solved.

8.2 pH Calculations

The unit pH is a most important term when dealing with acid-base chemistry because it describes to us the amount of H^+ ions that are present (and potentially reactive) in an aqueous solution.

When HCl is added to H_2O, the following reaction occurs:

$$HCl(aq) + H_2O(l) \square\ H^+(aq) + Cl^-(aq)$$

Instead of simply writing H^+ as one of the products of this reaction, we instead write $[H^+]$. As your text mentions, the production and concentration of $[H^+]$ is controlled by the dissociation of water. Water is capable of producing $[H^+]$ according to:

$$H_2O(aq) \square\ H^+_{(aq)} + OH^-(aq) \qquad K_w = 1.0 \times 10^{-14}$$

A solution consisting only of water will have a H^+ concentration of 1×10^{-7} M. If a strong acid such as HCl is added to the water, the $[H^+]$ will increase.

Instead of always trying to remember the hydrogen ion concentration of an acid solution as a value expressed to a power, chemists have simplified the values through the concept of pH. By definition:

$$pH = -\log[H^+]$$

If you know the H^+ equilibrium concentration, the pH may be obtained by entering the value of $[H^+]$ into your calculator, pressing the [log] key, and then the [+/-] to make the value positive. For example, if the $[H^+]$ is found to be 3.0×10^{-5} M, then the pH is calculated as:

$$pH = -\log [3.0 \times 10^{-5}]$$

$$pH = -(-4.52)$$

$$pH = 4.52$$

Important: Please note that there are two types of logarithmic functions on your calculator, [log] and [ln]. The pH definition is based on *log* and not *ln*.

We can check the pH calculations with a given $[H^+]$ by using the *log* relationship of powers to pH.

$$[H^+] = 1.0 \times 10^{-3}$$
$$pH = -\log [1.0 \times 10^{-3}]$$
$$pH = -(-3.00)$$
$$pH = 3.00$$

By similar reasoning, if $[H^+] = 1.0 \times 10^{-5}$, then the pH = 5.00. For the first example worked in this section where $[H^+] = 3.0 \times 10^{-5}$ M, we can predict that the pH should lie between 4 and 5. This second mental check can be of aid in confirming the answer from our calculator.

Another calculation that will be used in these studies is the determination of the $[H^+]$ if we are given the pH. This calculation can be performed by using the concept of *anti-logs*. From our earlier definition,

$$pH = -\log[H^+]$$
$$-pH = \log[H^+] \quad \text{(multiplying both sides by -1)}$$
$$10^{-pH} = 10^{\log[H^+]} \quad \text{(using the anti-log relationship)}$$
$$10^{-pH} = [H^+]$$
$$[H^+] = 10^{-pH}$$

To perform these calculations, use the $[y^x]$ or the $[10^x]$ function on your calculator. If the pH is 5.5, then:

$$[H^+] = 10^{-5.5}$$

$$[H^+] = 3.2 \times 10^{-6}\text{ M}$$

Note that several problems involve the calculation of the pH if you are given the OH^- concentration. The relationship between H^+ and OH^- is discussed in your textbook. From this relationship, we know that:

$$pK_w = [H^+] \times [OH^-]$$

and with the K_w of water at 25°C being 1×10^{-14}

$$14 = [H^+] \times [OH^-]$$

By knowing the $[OH^-]$, you can calculate the $[H^+]$ and then the pH.

Of course, you can also use the mathematics introduced in this chapter to calculate the pOH and relate it to the pK_w and the pH. By defining the pOH as the $-\log[OH^-]$, we can write the following expression:

$$pK_w = pH + pOH$$

$$14 = pH + pOH$$

This expression gives us the flexibility to calculate the pOH of a solution (when we are given the concentration of a base, such as NaOH), and translate it to the pH. Examine Problems 7 and 9 of the Self-Test for work with log calculations.

An important consideration of acid-base chemistry involves the strength of the compound. This relative strength of an acid or base can be described through the concept of electrolytes. For instance, the acid HCl is classified as a strong electrolyte and hence, a strong acid. The acid will completely dissociate in water; one mole of HCl yields one mole of H^+ and one mole of Cl^-. The presence of these ions in water causes the solution to conduct electricity. Therefore, it is called an electrolyte.

Other compounds may dissociate only to a small degree. If a mole of acetic acid is added to water, only about 5% of the acid will dissociate. This limited dissociation does not allow enough ions to be present in the solution in order for the solution to conduct electricity. Hence, acetic acid is called a weak electrolyte.

When you are calculating the pH of an acid or base solution, you will need to determine whether or not the acid (or base) involved in the reaction is a strong or weak species. As mentioned in your text, the $[H^+]$ for strong acids is usually the initial concentration of the reactants (*i.e.,* HCl) because of complete dissociation, whereas the $[H^+]$ for weak acids is determined from equilibrium calculations (as discussed in another chapter of your text).

Chemical Equilibrium

Goals: To calculate concentrations of products and reactants at equilibrium, and equilibrium constants.
Skills: Cross multiplication, powers, and roots.

Chemical equilibrium is based upon having a known stoichiometric chemical reaction in which the chemical expression can be written. With this information, it is possible to predict the yield of a chemical reaction. As an example, consider the chemical reaction

$$A + B \rightarrow 2C$$

The equilibrium expression for the reaction is:

$$K_{eq} = \frac{[C]^2}{[A][B]}$$

If you are given the equilibrium concentrations of A, B, and C, the equilibrium constant can be determined by entering these concentrations into the equation. For instance, if [A] = 0.10 M, [B] = 0.20 M, and [C] = 0.50 M, we can write:

$$K_{eq} = \frac{[0.50]^2}{[0.10][0.20]}$$

And

$$K_{eq} = 13$$

The calculation of K_{eq} involves mathematics very similar to Problem 7 of the Self-Test.

A variation to this problem may involve the determination of [C] if [A], [B], and K_C are given to you. If $K_C = 13$ with [A] = 0.50 M and [B] = 0.25, let's arrange the above equation to solve for [C].

$$K_C = \frac{[C]^2}{[0.50]\,[0.25]}$$

Next, each side of the equation is multiplied by the denominator, [0.50][0.25]. This act leaves only the term $[C]^2$ on the right side of the equation.

$$13 \times (\,[0.50][0.25]\,) = [C]^2$$

By rearranging and multiplying the numbers, we generate:

$$\sqrt{[C]^2} = \sqrt{1.6}$$

$$[C] = 1.3$$

The equilibrium concentration of C is 1.3 M.

Nuclear Chemistry

Goals: To determine the half-life of a radioactive element.
Skills: Multiplication, division, logarithms, and scientific notation.

In this chapter you will work problems concerning the decay of radioactive elements. This decay process follows first-order kinetics. The basic equation is:

$$\ln \frac{N_t}{N_0} = -kt \qquad \text{(decay equation)}$$

where N_0 is the amount of a radioisotope originally contained in sample, N_t is the amount contained at a specific time t, and k is the rate constant. From this equation, the expression for the half-life of a radioisotope can be determined. The half-life is the time required for 50% (one-half) of the radioisotope to decay.

The following example shows this relationship. If you had a vial containing 100 mg of an isotope used for medical imaging, how long would it take for only 50 mg of the isotope to remain active? Here, the value of N_0 would be 100 mg and N_t would be 50 mg.

$$\ln\left(\frac{50mg}{100mg}\right) = -kt_{1/2} \qquad \text{writing in the numbers}$$

$$\ln\left(\frac{1}{2}\right) = -kt_{1/2} \qquad \text{simplifying the terms}$$

$$-0.693 = -kt_{1/2} \quad \text{taking the ln of the left side}$$

$$0.693 = kt_{1/2} \quad \text{multiplying both sides by -1}$$

$$t_{1/2} = \frac{0.693}{k} \quad \text{rearranging for } t_{1/2}$$

We have not yet solved the problem. The equation above still needs a value for k in order to solve for $t_{1/2}$, the half-life. If the isotope has a half-life of 30.0 minutes, what is k? We can find it using the following equation:

$$k = \frac{0.693}{30.0\,\text{min}}$$

$$k = 0.0231\ \text{min}^{-1}$$

The decay rate constant for this isotope is 0.0231 min^{-1}. (Note that the units are expressed as inverse time, which cancels out the time units of $t_{1/2}$).

Often, isotope problems will ask you to find the amount of material still active after a certain decay period. If you know the half-life of the isotope, you can estimate this time based on the length of the decay time, expressed in lengths of half-life. (For the problem above, a decay time of 60 minutes would be two half-lives, or 1/2 times 1/2, which is 1/4; at 60 minutes only 1/4 of the isotope would remain.)

Let's look at a situation where the time involved is not a whole number ratio of the half-life. How much of the 100 mg sample of the tracer would still be active after 45.0 minutes? With this value of k and N_0 of 100 mg, we can use the decay equation to solve for the remaining mass of the tracer at $t = 45.0$ min, $N_{45.0}$:

$$\ln \frac{N_{45.0}}{100mg} = -(0.0231\ \text{min}^{-1}) \times (45.0\ \text{min})$$

$$\ln \frac{N_{45.0}}{100mg} = -1.0395$$

$$\frac{N_{45.0}}{100mg} = e^{-1.0395}\ \text{(using } e^x \text{ to remove ln x)}$$

$$\frac{N_{45.0}}{100mg} = 0.3536$$

$$N_{45.0} = 35.36\ \text{mg}$$

Therefore, after 45 minutes, only 35.36 mg of the isotope would remain.

The mathematics that are used in half-life problems are found in Problem 4 of the Self-Test. See Problem 5 for work with the mathematics used in carbon dating.

Self Test for Math Skills

Listed below are a series of questions designed to check your math skills. Please take no more than 30 minutes in answering them. You will find the answers in the back of this booklet.

1. How many significant figures are in the number 0.101?

2. Solve for x in the following equation with care to express the correct number of significant figures.

$$x = 12.011 + 1.00797$$

3. Solve for x in the following equation with care to express the correct number of significant figures.

$$x = \frac{1.057}{10.3}$$

4. Express the number 0.000356 in scientific notation.

5. Determine the mean, $\bar{x}$, for the following data: 25.12, 25.29, 24.95, 25.05 and 25.55.

6. Solve the equation: $x = 2.0 (\sin 35°)$.

7. Solve the equation: $0.405 = \cos(x)$.

8. Determine the value of x in the following equation.

$$(20.3) \times (0.0171) = \left(\frac{x}{60.3}\right) \times \left(1.35 \times 10^{3}\right)$$

9. What percentage of 15.5 is 1.25?

10. Solve the following equation for x.

$$\frac{1}{x} = \frac{4.3}{6.8}$$

11. Solve the following equation for x.

$$\frac{0.57}{0.22} = \frac{1.2}{x^2}$$

12. Calculate the value of x, where $x = (6.23 \times 1.22)^3$.

13. What is the fourth root of 17?

14. Solve the following equation for x.

$$x = 8.3 \ \square \left(\frac{1}{2^2} - \frac{1}{4^2}\right)$$

15. Determine the value of x in the following equation.

$$2.52 = \sqrt{\frac{1}{x}}$$

16. Calculate the value of x, where $x = \log(0.017)$.

17. Find the value of x, where $x = e^{\ln(12)}$.

18. Determine the value of x, where $1.53 = \log(x)$.

19. Solve for x in the following equation.

$$x = \frac{-1}{0.030} \ \square \ \ln\left(\frac{0.20}{0.50}\right)$$

20. Solve for x in the following equation.

$$x = \left(1.00 \square 10^3\right) e^{\left(\frac{-0.0220}{1.33}\right)}$$

Answers to Self Test For Math Skills

Answer		If you missed this problem, refer to:
1.	3	Sections 1.4 and 2.1
2.	13.019	Sections 1.3, 1.4, 1.6 and 2.1
3.	0.103	Sections 1.4 and 2.1
4.	3.56×10^{-4}	Sections 1.8, 1.10 and 2.2
5.	25.19	Section 2.3
6.	1.1	Section 1.2
7.	66.1°	Section 1.2
8.	0.0155	Sections 1.8, 1.9, 1.10 and 2.2
9.	8.06%	Sections 2.2 and 5
10.	1.6	Sections 1.2 and 6.1
11.	0.68	Section 1.2
12.	439	Sections 1.2, 1.8, 1.9 and 1.10
13.	2.03	Section 1.2
14.	1.6	Section 1.2
15.	0.157	Section 1.2
16.	-1.77	Sections 1.2 and 8.2
17.	12	Section 1.2
18.	34	Sections 1.2 and 8.2
19.	31	Sections 1.2 and 10
20.	984	Section 1.2